This Book Belongs To:

____

HERD OVERVIEW

Animal/ID	Tag#	Breed	Sex	Birth Date	Location/ Pasture	Pg#

HERD OVERVIEW

Animal/ID	Tag#	Breed	Sex	Birth Date	Location/ Pasture	Pg#

HERD OVERVIEW

Animal/ID	Tag#	Breed	Sex	Birth Date	Location/ Pasture	Pg#

HERD OVERVIEW

Animal/ID	Tag#	Breed	Sex	Birth Date	Location/ Pasture	Pg#

HERD NOTES

HERD NOTES

INDIVIDUAL ANIMAL HEALTH PROFILE

Animal Name/ID: ______________

Tag #: ________________________

Breed: ________________________

Sex: ___________________________

Birth Date: ____________________

Color/Markings: _______________

Sire (Father): ______________________

Dam (Mother): ____________________

Purchase Date: ___________________

Purchase Source: _________________

Current Pasture: __________________

GENERAL HEALTH NOTES

INDIVIDUAL ANIMAL HEALTH PROFILE

Animal Name/ID: ______________

Tag #: __________________________

Breed: _________________________

Sex: ____________________________

Birth Date: _____________________

Color/Markings: _______________

Sire (Father): ______________________

Dam (Mother): _____________________

Purchase Date: ____________________

Purchase Source: __________________

Current Pasture: __________________

GENERAL HEALTH NOTES

INDIVIDUAL ANIMAL HEALTH PROFILE

Animal Name/ID: ______________

Tag #: ______________

Breed: ______________

Sex: ______________

Birth Date: ______________

Color/Markings: ______________

Sire (Father): ______________

Dam (Mother): ______________

Purchase Date: ______________

Purchase Source: ______________

Current Pasture: ______________

GENERAL HEALTH NOTES

INDIVIDUAL ANIMAL HEALTH PROFILE

Animal Name/ID: ______________

Tag #: __________________________

Breed: _________________________

Sex: ____________________________

Birth Date: _____________________

Color/Markings: ________________

Sire (Father): _______________________

Dam (Mother): _____________________

Purchase Date: ____________________

Purchase Source: __________________

Current Pasture: ___________________

GENERAL HEALTH NOTES

INDIVIDUAL ANIMAL HEALTH PROFILE

Animal Name/ID: ______________

Tag #: ______________

Breed: ______________

Sex: ______________

Birth Date: ______________

Color/Markings: ______________

Sire (Father): ______________

Dam (Mother): ______________

Purchase Date: ______________

Purchase Source: ______________

Current Pasture: ______________

GENERAL HEALTH NOTES

INDIVIDUAL ANIMAL HEALTH PROFILE

Animal Name/ID: ________________

Tag #: ______________________________

Breed: ____________________________

Sex: _________________________________

Birth Date: _______________________

Color/Markings: _________________

Sire (Father): __________________________

Dam (Mother): _______________________

Purchase Date: ______________________

Purchase Source: ____________________

Current Pasture: _____________________

GENERAL HEALTH NOTES

INDIVIDUAL ANIMAL HEALTH PROFILE

Animal Name/ID: ______________

Tag #: ______________

Breed: ______________

Sex: ______________

Birth Date: ______________

Color/Markings: ______________

Sire (Father): ______________

Dam (Mother): ______________

Purchase Date: ______________

Purchase Source: ______________

Current Pasture: ______________

GENERAL HEALTH NOTES

INDIVIDUAL ANIMAL HEALTH PROFILE

Animal Name/ID: ______________

Tag #: ______________

Breed: ______________

Sex: ______________

Birth Date: ______________

Color/Markings: ______________

Sire (Father): ______________

Dam (Mother): ______________

Purchase Date: ______________

Purchase Source: ______________

Current Pasture: ______________

GENERAL HEALTH NOTES

INDIVIDUAL ANIMAL HEALTH PROFILE

Animal Name/ID: ______________

Tag #: ________________________

Breed: ________________________

Sex: __________________________

Birth Date: ____________________

Color/Markings: _______________

Sire (Father): _________________________

Dam (Mother): ______________________

Purchase Date: _____________________

Purchase Source: ___________________

Current Pasture: ___________________

GENERAL HEALTH NOTES

INDIVIDUAL ANIMAL HEALTH PROFILE

Animal Name/ID: ______________

Tag #: __________________________

Breed: __________________________

Sex: ____________________________

Birth Date: ______________________

Color/Markings: _______________

Sire (Father): _________________________

Dam (Mother): _______________________

Purchase Date: _____________________

Purchase Source: __________________

Current Pasture: ___________________

GENERAL HEALTH NOTES

INDIVIDUAL ANIMAL HEALTH PROFILE

Animal Name/ID: ______________

Tag #: __________________________

Breed: _________________________

Sex: ____________________________

Birth Date: _____________________

Color/Markings: _______________

Sire (Father): ______________________

Dam (Mother): ____________________

Purchase Date: ___________________

Purchase Source: _________________

Current Pasture: _________________

GENERAL HEALTH NOTES

INDIVIDUAL ANIMAL HEALTH PROFILE

Animal Name/ID: ______________

Sire (Father): ______________

Tag #: ______________

Dam (Mother): ______________

Breed: ______________

Purchase Date: ______________

Sex: ______________

Purchase Source: ______________

Birth Date: ______________

Current Pasture: ______________

Color/Markings: ______________

GENERAL HEALTH NOTES

INDIVIDUAL ANIMAL HEALTH PROFILE

Animal Name/ID: ______________

Tag #: __________________________

Breed: _________________________

Sex: ____________________________

Birth Date: ____________________

Color/Markings: _______________

Sire (Father): ______________________

Dam (Mother): ____________________

Purchase Date: ___________________

Purchase Source: _________________

Current Pasture: __________________

GENERAL HEALTH NOTES

INDIVIDUAL ANIMAL HEALTH PROFILE

Animal Name/ID: ____________

Tag #: ____________

Breed: ____________

Sex: ____________

Birth Date: ____________

Color/Markings: ____________

Sire (Father): ____________

Dam (Mother): ____________

Purchase Date: ____________

Purchase Source: ____________

Current Pasture: ____________

GENERAL HEALTH NOTES

INDIVIDUAL ANIMAL HEALTH PROFILE

Animal Name/ID: ______________

Tag #: __________________________

Breed: __________________________

Sex: ____________________________

Birth Date: ______________________

Color/Markings: ________________

Sire (Father): _______________________

Dam (Mother): _____________________

Purchase Date: ____________________

Purchase Source: _________________

Current Pasture: __________________

GENERAL HEALTH NOTES

INDIVIDUAL ANIMAL HEALTH PROFILE

Animal Name/ID: ______________

Tag #: ______________

Breed: ______________

Sex: ______________

Birth Date: ______________

Color/Markings: ______________

Sire (Father): ______________

Dam (Mother): ______________

Purchase Date: ______________

Purchase Source: ______________

Current Pasture: ______________

GENERAL HEALTH NOTES

INDIVIDUAL ANIMAL HEALTH PROFILE

Animal Name/ID: ____________

Tag #: ____________

Breed: ____________

Sex: ____________

Birth Date: ____________

Color/Markings: ____________

Sire (Father): ____________

Dam (Mother): ____________

Purchase Date: ____________

Purchase Source: ____________

Current Pasture: ____________

GENERAL HEALTH NOTES

INDIVIDUAL ANIMAL HEALTH PROFILE

Animal Name/ID: ______________

Tag #: ______________

Breed: ______________

Sex: ______________

Birth Date: ______________

Color/Markings: ______________

Sire (Father): ______________

Dam (Mother): ______________

Purchase Date: ______________

Purchase Source: ______________

Current Pasture: ______________

GENERAL HEALTH NOTES

INDIVIDUAL ANIMAL HEALTH PROFILE

Animal Name/ID: ____________

Tag #: ____________

Breed: ____________

Sex: ____________

Birth Date: ____________

Color/Markings: ____________

Sire (Father): ____________

Dam (Mother): ____________

Purchase Date: ____________

Purchase Source: ____________

Current Pasture: ____________

GENERAL HEALTH NOTES

INDIVIDUAL ANIMAL HEALTH PROFILE

Animal Name/ID: ________________

Tag #: ______________________________

Breed: ____________________________

Sex: _______________________________

Birth Date: ______________________

Color/Markings: _________________

Sire (Father): __________________________

Dam (Mother): _______________________

Purchase Date: ______________________

Purchase Source: ____________________

Current Pasture: _____________________

GENERAL HEALTH NOTES

MEDICAL TREATMENTS & EVENTS

Date	Diagnosis /Issue	Treatment/ Medication	Notes

FOLLOW-UP/OUTCOME NOTES

TREATMENT LOG

Date	Animal ID	Symptoms/ Condition	Diagnosis	Medication	Dosage	Notes

TREATMENT LOG

Date	Animal ID	Symptoms/ Condition	Diagnosis	Medication	Dosage	Notes

TREATMENT LOG

Date	Animal ID	Symptoms/ Condition	Diagnosis	Medication	Dosage	Notes

TREATMENT LOG

Date	Animal ID	Symptoms/ Condition	Diagnosis	Medication	Dosage	Notes

TREATMENT LOG

Date	Animal ID	Symptoms/ Condition	Diagnosis	Medication	Dosage	Notes

TREATMENT LOG

Date	Animal ID	Symptoms/ Condition	Diagnosis	Medication	Dosage	Notes

TREATMENT LOG

Date	Animal ID	Symptoms/ Condition	Diagnosis	Medication	Dosage	Notes

TREATMENT LOG

Date	Animal ID	Symptoms/ Condition	Diagnosis	Medication	Dosage	Notes

TREATMENT LOG

Date	Animal ID	Symptoms/ Condition	Diagnosis	Medication	Dosage	Notes

TREATMENT LOG

Date	Animal ID	Symptoms/ Condition	Diagnosis	Medication	Dosage	Notes

TREATMENT LOG

Date	Animal ID	Symptoms/ Condition	Diagnosis	Medication	Dosage	Notes

TREATMENT LOG

Date	Animal ID	Symptoms/ Condition	Diagnosis	Medication	Dosage	Notes

TREATMENT LOG

Date	Animal ID	Symptoms/ Condition	Diagnosis	Medication	Dosage	Notes

TREATMENT LOG

Date	Animal ID	Symptoms/ Condition	Diagnosis	Medication	Dosage	Notes

TREATMENT LOG

Date	Animal ID	Symptoms/ Condition	Diagnosis	Medication	Dosage	Notes

TREATMENT LOG

Date	Animal ID	Symptoms/ Condition	Diagnosis	Medication	Dosage	Notes

TREATMENT LOG

Date	Animal ID	Symptoms/ Condition	Diagnosis	Medication	Dosage	Notes

TREATMENT LOG

Date	Animal ID	Symptoms/ Condition	Diagnosis	Medication	Dosage	Notes

TREATMENT LOG

Date	Animal ID	Symptoms/ Condition	Diagnosis	Medication	Dosage	Notes

TREATMENT LOG

Date	Animal ID	Symptoms/ Condition	Diagnosis	Medication	Dosage	Notes

TREATMENT LOG

Date	Animal ID	Symptoms/ Condition	Diagnosis	Medication	Dosage	Notes

TREATMENT LOG

Date	Animal ID	Symptoms/ Condition	Diagnosis	Medication	Dosage	Notes

TREATMENT LOG

Date	Animal ID	Symptoms/ Condition	Diagnosis	Medication	Dosage	Notes

TREATMENT LOG

Date	Animal ID	Symptoms/ Condition	Diagnosis	Medication	Dosage	Notes

TREATMENT LOG

Date	Animal ID	Symptoms/ Condition	Diagnosis	Medication	Dosage	Notes

TREATMENT LOG

Date	Animal ID	Symptoms/ Condition	Diagnosis	Medication	Dosage	Notes

TREATMENT LOG

Date	Animal ID	Symptoms/ Condition	Diagnosis	Medication	Dosage	Notes

TREATMENT LOG

Date	Animal ID	Symptoms/ Condition	Diagnosis	Medication	Dosage	Notes

TREATMENT LOG

Date	Animal ID	Symptoms/ Condition	Diagnosis	Medication	Dosage	Notes

TREATMENT LOG

Date	Animal ID	Symptoms/ Condition	Diagnosis	Medication	Dosage	Notes

VACCINATION LOG

Date	Animal ID	Dosage	Vaccine	Notes

VACCINATION LOG

Date	Animal ID	Dosage	Vaccine	Notes

VACCINATION LOG

Date	Animal ID	Dosage	Vaccine	Notes

VACCINATION LOG

Date	Animal ID	Dosage	Vaccine	Notes

VACCINATION LOG

Date	Animal ID	Dosage	Vaccine	Notes

VACCINATION LOG

Date	Animal ID	Dosage	Vaccine	Notes

VACCINATION LOG

Date	Animal ID	Dosage	Vaccine	Notes

VACCINATION LOG

Date	Animal ID	Dosage	Vaccine	Notes

VACCINATION LOG

Date	Animal ID	Dosage	Vaccine	Notes

VACCINATION LOG

Date	Animal ID	Dosage	Vaccine	Notes

DEWORMING LOG

Date	Animal ID	Product	Dosage	Next Due Date	Notes

DEWORMING LOG

Date	Animal ID	Product	Dosage	Next Due Date	Notes

DEWORMING LOG

Date	Animal ID	Product	Dosage	Next Due Date	Notes

DEWORMING LOG

Date	Animal ID	Product	Dosage	Next Due Date	Notes

DEWORMING LOG

Date	Animal ID	Product	Dosage	Next Due Date	Notes

DEWORMING LOG

Date	Animal ID	Product	Dosage	Next Due Date	Notes

DEWORMING LOG

Date	Animal ID	Product	Dosage	Next Due Date	Notes

DEWORMING LOG

Date	Animal ID	Product	Dosage	Next Due Date	Notes

DEWORMING LOG

Date	Animal ID	Product	Dosage	Next Due Date	Notes

DEWORMING LOG

Date	Animal ID	Product	Dosage	Next Due Date	Notes

GENERAL NOTES

GENERAL NOTES

GENERAL NOTES

GENERAL NOTES

GENERAL NOTES

GENERAL NOTES

GENERAL NOTES

GENERAL NOTES

GENERAL NOTES

GENERAL NOTES

GENERAL NOTES

GENERAL NOTES

GENERAL NOTES

GENERAL NOTES

GENERAL NOTES

GENERAL NOTES

GENERAL NOTES

GENERAL NOTES

GENERAL NOTES

GENERAL NOTES

GENERAL NOTES

Made in the USA
Middletown, DE
04 November 2026

36544881R00057